YOUR KNOWLEDGE HAS VALUE

- We will publish your bachelor's and master's thesis, essays and papers

- Your own eBook and book - sold worldwide in all relevant shops

- Earn money with each sale

Upload your text at www.GRIN.com and publish for free

Nanotechnology and How it Works. Possible Ways to Use and Apply Nanotechnology

Rajni Garg

Bibliographic information published by the German National Library:

The German National Library lists this publication in the National Bibliography; detailed bibliographic data are available on the Internet at http://dnb.dnb.de.

ISBN: 9783346720627
This book is also available as an ebook.

Print and binding: Books on Demand GmbH, Norderstedt, Germany
Printed on acid-free paper from responsible sources.

The present work has been carefully prepared. Nevertheless, authors and publishers do not incur liability for the correctness of information, notes, links and advice as well as any printing errors.

GRIN web shop: https://www.grin.com/document/1272017

Nanotechnology: The big scenario

Rajni Garg[a]

[a]R&D Department, Institute of Sci-Tech affairs, Mohali, Punjab, India

Table of Contents

1. Nanotechnology 2

2. What can be said for and against 4

3.What is nanotechnology and how it works? 5

4. Possible Ways to Use and Apply Nanotechnology...5

5. Risk assessment 6

6. Limitations put on the Scope 8

7. Conclusion and concerns for the future 10

References..13

1. Nanotechnology

Nanotechnology is the branch of science and engineering that focuses on designing, making, and using frameworks, systems, and structures that are made by controlling atoms and molecules at the nanoscale (Rajni Garg et al. 2022). This means that different structures, devices, and systems must have at least one dimension that is 100 nanometers (100 millionths of a millimetre) or smaller. Or, you could say that the size must be less than 100 nanometers (Sharma, Garg, and Kumari 2020). There are many natural structures with sizes of one or more nanometers. Nanostructures have been accidentally used in a wide range of applications since the beginning of technology (Foroughi et al. 2021). However, it wasn't until fairly recently that it became possible to make these structures on purpose. When compared to the same materials made on a larger scale, nanotechnology products have very different properties and effects (Caon, Martelli, and Fakhouri 2017). One of the most common ways nanotechnology is used is to make new materials that are different in these ways. This is because, compared to bigger particles, nanoparticles have a very high ratio of surface area to volume (Hasheminya and Dehghannya 2020). Other effects can be seen at this scale but not at larger scales. These effects can be seen. In this opinion, the different nanotechnology-related terms are used in a way that is consistent with the following definitions for the most important general terms: nanoparticle, nanomaterial, nanoscale, nanoscale device, nanoscale material, nanoscale device, nanoscale device, nanoscale device, nanoscale device, nanoscale device, nanoscale device, nanoscale

• Something is on the nanoscale if it has one or more dimensions that are 100 nanometers or smaller (Ghosh et al. 2022).

• The study of what occurs and how things can be altered at the atomic, molecule, and macromolecular sizes, where things are significantly different from larger scales (Patra et al. 2018).

• The study of what happens at the atomic, molecular, and macromolecular levels and how things can change at those levels (Rajni Garg et al. 2022).

•Nanoscience is the study of things that happen at the nanoscale level (Sozer and Kokini 2009).

• "Nanotechnology" refers to the process of designing, analysing, making, and using structures, devices, and systems that are made by changing their shape and size on the nanoscale. The term "nanotechnology" also refers to the process (S. Kumar et al. 2019).

• A type of material is called a "nanomaterial" if it has one or more outer dimensions or an inner structure and could have different properties than the same type of material that doesn't have nanoscale features (V. Kumar and Arora 2020).

•A nanoparticle is a type of particle that has at least one size that can be measured on the nanoscale. (It's important to note that in this study, nanoparticles are thought to have two or more dimensions when viewed on the nanoscale.) (Rishav Garg and Garg 2020)

•A nanocomposite is a type of composite in which at least one of the phases has at least one dimension that can be measured on the nanoscale. This sort of composite is referred to by the name "nanocomposite." (Naseem and Durrani 2021)

•Something is said to be "nanostructured" if it has a structure on the nanoscale (Holzinger, Goff, and Cosnier 2014).

2. What can be said for and against

Nanotechnology's uses could have a big effect on society and help people in many ways. Nanotechnology is now used in many business areas, including information and communications, food technology, energy technology, and especially medical products and therapies (Bhardwaj, Lata, and Garg 2022). Also, using nanomaterials could lead to the discovery of new ways to reduce the harmful effects of pollution on the environment (Eddy et al. 2022). On the other hand, these new compounds could pose risks to human health that we don't know about yet. Humans have come up with a number of ways to protect themselves from a wide range of environmental threats, which can be of different strengths (Mohaghegh et al. 2020). On the other hand, they had never worked with synthetic nanoparticles or known what characteristics they had before not too long ago. So, it's possible that the body's normal defences, like the immune and inflammatory systems, might not be able to fight these nanoparticles as well as they could (Mittal, Chisti, and Banerjee 2013). Because nanoparticles are so small, this is the case. Nanoparticles might also move around and stay in the environment, which means they might have an effect on the natural world (Rajni Garg, Kumari, et al. 2021). When thinking about the risks nanostructures might pose to a person's health, there are two different kinds of nanostructures to think about:

• those in which the structure is a free particle in and of itself, which are called "free nanoparticles" and are the most worrying; and • those in which the structure is a free particle in and of itself (Shukla et al. 2009).

• Those in which the nanostructure is a natural part of a larger object, like materials with nanomaterial coatings. These are different from other nanostructured materials (Vipan et al. 2018).

But as long as the nanoparticles are attached to the carrier, there is no reason to think that they pose a greater risk to human health or the environment than materials that are bigger. This is true no matter if the nanoparticles are thought to be dangerous or not (Asimbaya et al. 2022).

3.What is nanotechnology and how it works?

Nanotechnologies are the fields of science and engineering that use things that happen on the nanometer scale to design, characterise, make, and use new materials, structures, devices, and systems. The field that includes all of these subfields is called "nanotechnology." (Rajni Garg, Kumari, et al. 2021) Even though there are many nanometer-sized structures in nature, such as molecules in the human body and food ingredients, and even though many technologies have used nanoscale structures for a long time, it has only been possible in the last 25 years to change molecules and structures on purpose (Drexler 2006). Because of recent improvements in nanotechnology, scientists can now make new materials with properties that weren't possible before. Nanotechnology is different from other fields of technology because it can control things on a scale that is on the nanoscale (Díez-Pascual and Díez-Vicente 2014).

4. Possible Ways to Use and Apply Nanotechnology

There is no doubt that the many ways nanotechnology can be used have the potential to have a huge effect on society. This is something that can be said for sure. It is reasonable to think that, in the long run, using nanotechnology will be very helpful to both people and organisations in a wide range of settings (Huang et al. 2018). At these sizes, the ratio of surface area to volume is very high. This, along with the quantum effects that aren't visible at larger scales, is what causes new things to happen. A lot of these applications require the creation of brand-new materials that, because they can work at the nanoscale, can do things that no other materials can (Bhatia 2019). Some examples of these types of materials are very thin films, which are used in the fields of catalysis and electronics, two-dimensional nanotubes and nanowires, which are used in the fields of optical and magnetic systems, and nanoparticles, which are used in the fields of cosmetics, drugs, and coatings (Rishav Garg, Garg, Okon Eddy, et al. 2022). The industries that are most open to nanotechnology are those that deal with information and communications, which includes fields like electronics and optoelectronics, food technology, energy technology, and medical goods. Some other businesses are: In fields like pharmaceuticals, drug delivery systems, diagnostics, and medical technology, where the terms "nanomedicine" and "bionanotechnology" are already common, there are many different ways that these terms can be used. When it comes to making the environment less polluted, products made with nanotechnology may bring up new problems that need to be solved (Vazquez-Munoz and Lopez-Ribot 2020).

5. Risk assessment

But just as the things that happen at the nanoscale may be very different from those that happen at larger scales and may be useful to humans, these newly discovered processes and their products may pose new health risks to humans and the environment as a whole. These risks may involve very different ways of messing with the way humans and other species' bodies work (Oves, Khan, and Ismail 2017).

Also, both the ways these things are done and the things they make may cause health problems for the same group of people. These possibilities may very well be about what happens to free nanoparticles that are made by nanotechnology and then released into the environment on purpose or by accident, or that are delivered directly to people through the use of a nanotechnology-based product (Rishav Garg, Garg, Khan, et al. 2022). Because of their size, shape, or make-up, these nanoparticles could be harmful to people. The medical community should be particularly concerned about the people whose employment put them in continuous and extended contact with free nanoparticles. Because of their jobs, these people have to deal with free nanoparticles (Pozdnyakov et al. 2016). The process is based on the types of agents that are most common, and size is an important factor in this case. As a result of evolution, humans have learned to protect themselves from both living and dead environmental agents. This is at the heart of the worries that have been raised about the potential for harmful health effects (Rajni Garg, Rani, et al. 2021). When nanoparticles with never-before-seen properties are around, it's possible that the body's normal defence systems, like the immune and inflammatory systems, will be put to the test. This is because nanoparticles have properties that have never been seen before. Because of the products of nanotechnology, the ways that nanoparticles move around and stay in the environment may have an effect on the environment. There's no reason to say this can't happen (Rajni Garg et al. 2022).

If there is a chance that a completely new danger will be found, it is very important to look into the nature of the risk in great detail. The results of this study can then be used, if necessary, as part of the risk management process (R.D. Singh 2011). Most people agree that this is the best way to figure out how dangerous nanotechnology is. Many international organisations, such as the Asia Pacific Nanotechnology Forum in 2005, governmental bodies in the European Union, such as the European Commission in 2004, national institutions, such as the United States National Science and Technology Council in 2004, IEEE in 2004, and the United States National Institute of Environmental Health Sciences in 2004, non-governmental organisations, such as the UN-NGLS in 2005, and academic institutions have issued statements (He, Deng, and Hwang 2019).

The European Council has said that it is important to pay special attention to the possible risks of nanotechnology products throughout their entire life cycle. The European Commission has said that it wants to work with other countries to create a set of shared principles for the safe, responsible, sustainable, and socially acceptable use of nanotechnology (Mageswari et al.

2016). The European Council has also said that we need to pay extra attention to the possible risks of nanotechnology-based products throughout their whole life cycle. During the European Council's discussion of the issue, both of these points were brought up.

6. Limitations put on the Scope

There are many ways to understand both the word "nanotechnology" and the things that can be made with it. Most of the time, these things are made to do very specific jobs or reach very specific goals (Eddy and Garg 2022).

Because of this, it was decided that the basic scientific ideas behind nanotechnology were more important than the exact words used to describe it. As a result, the basic scientific ideas will be discussed first. Because of this, one might think that the nanoscale goes from the atomic level, which is about 0.2 nm in size, all the way up to about 100 nm. If the same materials are this small or this big, they can have very different qualities when compared to the same materials that are bigger. This is true not only because the ratio of surface area to mass is much higher, but also because quantum effects start to play a role at these scales. These effects cause big changes in a lot of different physical properties (Rajni Garg, Garg, and Okon Eddy 2022).

It is important to note that nanoscience and nanotechnology have grown quickly in the past few years, and that the language used in the fields that have contributed to this growth has not stayed the same. During this time, there have also been different ways to talk about things. Also, as this paper points out, measuring the parameters at the nanoscale accurately has been and still is a big challenge (Akintelu et al. 2020). Because of these problems, it is not always possible to have full faith in the data collected and the conclusions drawn about certain phenomena that are related to certain properties of nanostructures and nanomaterials. The reason for this is that these things are very specialised. Because the nanoscale is a very small scale, this is the case. Even though there may be mistakes and inconsistencies in the literature, this point of view admits that the situation is inevitable and has come to a few general conclusions as a result of its analysis (Adewuyi 2020). This point of view acknowledges that the situation couldn't have been avoided. Even though this opinion says that nanoscale should now be thought of as having dimensions up to 100 nm, it admits that some of the literature has talked about nanoscale as having dimensions bigger than 100 nm. Even though this opinion says that nanoscale should now be thought of as having sizes up to 100 nm, this opinion is not supported by evidence. This is because of how the word "nanoscale" is defined in this ruling. It says that it should now be thought of as including sizes up to 100 nm (Gandhi, Garg, and Eddy 2021). Most research on particles has called them either ultrafine, fine, or regular particles. This is because each of these groups describes a different size level. This is especially true of the research on aerosols, air pollution, and the harmful effects of breathing them in. This study is based on the idea that "ultrafine particles," unless it is made clear that their roles are different from nanoparticles, do the same things that nanoparticles do (Zhou et al. 2018).

When talking about nanoparticles, it is important to remember that a sample of a substance with nanoparticles will not be monodisperse, but will usually have particles of different sizes. Because nanoparticles come in many different sizes, this is the case (Rajni Garg 2011). This is a very important point that you shouldn't forget. Because of this, it is much harder to give an accurate assessment of the properties of the nanoscale, especially when the doses are taken into account for toxicological research. In this point of view, there are a lot of references to studies about particle exposure and toxicity data. When talking about these studies, we may talk about the particle size as a single number (like 40 nm) or a range (like 40–80 nm) from the articles, but it's important to remember that these are just rough estimates (Khan et al. 2016).

There is also a chance that the nanoparticles will tend to stick together under certain circumstances. This is something that should be looked into more. It is possible that a group of nanoparticles, which may have sizes that are measured in microns instead of nanometers, will act differently than the nanoparticles if they are looked at separately (Rajni Garg, Kumari, et al. 2021). On the other hand, there is no reason to think that the group will behave the same way as a single large particle. This guess might be based on the fact that the size of a group can be measured in microns instead of nanometers (Almomani et al. 2020). You can also expect that the way nanoparticles behave will depend on how easily they dissolve and how easily they break down. Neither the chemical composition nor the size of the particles is guaranteed to stay the same over time. This is because nanoparticles are so small that their ability to dissolve and their susceptibility to break down are built into them. This is because neither the chemical composition nor the size of the particles can be measured directly (Rai, Acharya, and Dey 2012).

7. Conclusion and concerns for the future

Keeping in mind the above definitions and qualifiers, it is clear that there are two different types of nanostructures that need to be taken into account in terms of their properties and possible health risks. We need to think about these nanostructures. The first kind of nanostructure is one in which the structure itself is a free particle. The second type of nanostructure is one where the nanostructure is a part of something bigger (Rajni Garg, Rani, et al. 2021).

This type of material includes things like nanocomposites and nanocrystalline solids. Nanocomposites are solid materials that have one or more phases that are dispersed as nanoscale particles. Nanocrystalline solids, on the other hand, are solids where each crystal has dimensions on the nanoscale. Solid materials include both nanocomposites and nanocrystalline solids. This category also includes things that have nanoscale-sized features on their surface topography, as well as functional parts that have important features that are measured in nanometers, most of which are electronic parts. This category also includes things that have nanoscale-sized features on their surface topography. By applying coatings made of nanosized parts, it is possible to change the surface in ways that are useful for medical applications. This opinion acknowledges that the materials and products listed above exist and that the properties of materials with sizes on the nanoscale can affect how they interact with biological systems (Rajni Garg 2012). This opinion also takes into account the fact that the properties of materials with sizes on the nanoscale can have an effect on biological systems. But, even though the science of interactions between biological systems and nanotopographical features is growing quickly, not much is known about the ways in which these interactions could cause problems. Even though the study of how biological systems and nanotopographical features interact is growing quickly, this is still the case. The risk would be directly related to how well the substance stuck to the carrier material, and it would be linked to how much of the substance leaked out while the product was being used or after it had served its purpose. As long as the nanomaterials are stuck to the surface of the carrier, there is no reason to think that immobilised nanoparticles pose a higher risk to human health or the environment than larger materials do. This is because nanoparticles that are stuck together are on the carrier's surface.

The first category, which includes free nanoparticles, is the one that worries people the most about possible health risks, so that's where this opinion will focus most of the time. Nanotubes are in the second group, which is the least dangerous to health. The word "free" needs to be qualified because it implies that the chemical in question is made up of tiny particles at some point during its production or use. When the material is used, the individual particles may be mixed with a certain amount of another substance, which could be a gas, a liquid, or a solid, usually to make a paste, a gel, or a coating. This can happen no matter if the other thing is a gas, a liquid, or a solid. Depending on what state it is in, this extra part could be a gas, a liquid, or a solid. Even though the type of phase in which these particles are dispersed will affect how bioavailable they are, you can still think of these particles as free even though you know that the type of phase in which they are dispersed will affect this. This group would include ultrafine aerosols and colloids, as well as cream-based cosmetics and drug preparations. Most of the current studies on the health risks of nanotechnology have focused on these kinds of products. This point of view is mostly worried about the risks that can come from making and using products that contain engineered nanomaterials. Nanostructures that come from living things, such as proteins, phospholipids, lipids, and the like, are not taken into account in this situation.

References

Adewuyi, Adewale. 2020. "Chemically Modified Biosorbents and Their Role in the Removal of Emerging Pharmaceuticalwaste in the Water System." *Water (Switzerland)* 12 (6): 1–31. https://doi.org/10.3390/W12061551.

Akintelu, Sunday Adewale, Aderonke Similoluwa Folorunso, Femi Adekunle Folorunso, and Abel Kolawole Oyebamiji. 2020. "Green Synthesis of Copper Oxide Nanoparticles for Biomedical Application and Environmental Remediation." *Heliyon* 6 (7): e04508. https://doi.org/10.1016/j.heliyon.2020.e04508.

Almomani, Fares, Rahul Bhosale, Majeda Khraisheh, Anand kumar, and Thakir Almomani. 2020. "Heavy Metal Ions Removal from Industrial Wastewater Using Magnetic Nanoparticles (MNP)." *Applied Surface Science* 506: 144924. https://doi.org/10.1016/j.apsusc.2019.144924.

Asimbaya, Christopher, Nelly Maria Rosas-Laverde, Salome Galeas, Alexis Debut, Victor H. Guerrero, and Alina Pruna. 2022. "Magnetite Impregnated Lignocellulosic Biomass for Zn(II) Removal." *Materials* 15 (3). https://doi.org/10.3390/ma15030728.

Bhardwaj, Sweta, Suman Lata, and Rajni Garg. 2022. "Application of Nanotechnology for Preventing Postharvest Losses of Agriproducts." *The Journal of Horticultural Science and Biotechnology* 00 (00): 1–14. https://doi.org/10.1080/14620316.2022.2091488.

Bhatia, S. C. 2019. "Nanotechnology in Agriculture and Food Industry." *Food Biotechnology* 16 (2): 363–77. https://doi.org/10.1201/9781315156491-23.

Caon, Thiago, Silvia Maria Martelli, and Farayde Matta Fakhouri. 2017. "New Trends in the Food Industry: Application of Nanosensors in Food Packaging." In *Nanobiosensors*, 773–804. Elsevier. https://doi.org/10.1016/b978-0-12-804301-1.00018-7.

Díez-Pascual, Ana M., and Angel L. Díez-Vicente. 2014. "ZnO-Reinforced Poly(3-

Hydroxybutyrate-Co-3-Hydroxyvalerate) Bionanocomposites with Antimicrobial

Function for Food Packaging." *ACS Applied Materials and Interfaces* 6 (12): 9822–34.

https://doi.org/10.1021/am502261e.

Drexler, K. Eric. 2006. "Engines of Creation 2.0." The Coming Era of Nanotechnology."

Anchor Books- Doubleday 1986: 576.

http://www1.appstate.edu/dept/physics/nanotech/EnginesofCreation2_8803267.pdf.

Eddy, Nnabuk Okon, and Rajni Garg. 2022. "CaO Nanoparticles." In *Handbook of Research*

on Green Synthesis and Applications of Nanomaterials, 247–68.

https://doi.org/10.4018/978-1-7998-8936-6.ch011.

Eddy, Nnabuk Okon, Rajni Garg, Rishav Garg, Augustine O. Aikoye, and Benedict I. Ita.

2022. "Waste to Resource Recovery: Mesoporous Adsorbent from Orange Peel for the

Removal of Trypan Blue Dye from Aqueous Solution." *Biomass Conversion and*

Biorefinery. https://doi.org/10.1007/s13399-022-02571-5.

Foroughi, Mohammad Mehdi, Shohreh Jahani, Zahra Aramesh-Boroujeni, Meisam

Rostaminasab Dolatabad, and Kiana Shahbazkhani. 2021. "Synthesis of 3D Cubic of

Eu3+/Cu2O with Clover-like Faces Nanostructures and Their Application as an

Electrochemical Sensor for Determination of Antiretroviral Drug Nevirapine." *Ceramics*

International 47 (14): 19727–36. https://doi.org/10.1016/j.ceramint.2021.03.311.

Gandhi, C. P., Rajni Garg, and Nnabuk Okon Eddy. 2021. "Application of Biosynthesized

Nano-Catalyst for Biodiesel Synthesis and Impact Assessment of Factors Influencing the

Yield." *Nanosystems: Physics, Chemistry, Mathematics* 12 (6): 808–17.

https://doi.org/10.17586/2220-8054-2021-12-6-808-817.

Garg, Rajni, Rishav Garg, and Nnabuk Okon Eddy. 2022. *Handbook of Research on Green*

Synthesis and Applications of Nanomaterials. Edited by Rajni Garg, Rishav Garg, and

Nnabuk Okon Eddy. Vol. i. Advances in Chemical and Materials Engineering. IGI

Global. https://doi.org/10.4018/978-1-7998-8936-6.

Garg, Rajni, Mamta Kumari, Mandeep Kumar, Sourabh Dhiman, and Rishav Garg. 2021.

"Green Synthesis of Calcium Carbonate Nanoparticles Using Waste Fruit Peel Extract."

Materials Today: Proceedings 46 (xxxx): 6665–68.

https://doi.org/10.1016/j.matpr.2021.04.124.

Garg, Rajni, Priya Rani, Rishav Garg, Mohammad Amir, Nadeem Ahmad, Afzal Husain, and

Juliana Heloisa. 2022. "Biomedical and Catalytic Applications of Agri-Based

Biosynthesized Silver Nanoparticles." *Environmental Pollution* 310 (June): 119830.

https://doi.org/10.1016/j.envpol.2022.119830.

Garg, Rajni, Priya Rani, Rishav Garg, and Nnabuk Okon Eddy. 2021. "Study on Potential

Applications and Toxicity Analysis of Green Synthesized Nanoparticles." *Turkish

Journal of Chemistry* 45 (6): 1690–1706. https://doi.org/10.3906/kim-2106-59.

Garg, Rishav, and Rajni Garg. 2020. "Effect of Zinc Oxide Nanoparticles on Mechanical

Properties of Silica Fume-Based Cement Composites." *Materials Today: Proceedings*

43: 778–83. https://doi.org/10.1016/j.matpr.2020.06.168.

Garg, Rishav, Rajni Garg, Md. Amir Khan, Manjeet Bansal, and Vinod Garg. 2022.

"Utilization of Biosynthesized Silica-Supported Iron Oxide Nanocomposites for the

Adsorptive Removal of Heavy Metal Ions from Aqueous Solutions." *Environmental

Science and Pollution Research*, no. 0123456789: 1–10.

https://doi.org/https://doi.org/10.21203/rs.3.rs-1394501/v1.

Garg, Rishav, Rajni Garg, Nnabuk Okon Eddy, Abdulaziz Ibrahim Almohana, Sattam Fahad

Almojil, Mohammad Amir Khan, and Seung Ho Hong. 2022. "Biosynthesized Silica-

Based Zinc Oxide Nanocomposites for the Sequestration of Heavy Metal Ions from

Aqueous Solutions." *Journal of King Saud University - Science* 34 (4): 101996. https://doi.org/10.1016/j.jksus.2022.101996.

Ghosh, Ujjyani, Khondakar Sayef Ahammed, Snehasis Mishra, and Asim Bhaumik. 2022. "The Emerging Roles of Silver Nanoparticles to Target Viral Life Cycle and Detect Viral Pathogens." *Chemistry - An Asian Journal* 17 (5). https://doi.org/10.1002/asia.202101149.

Hasheminya, Seyedeh Maryam, and Jalal Dehghannya. 2020. "Green Synthesis and Characterization of Copper Nanoparticles Using Eryngium Caucasicum Trautv Aqueous Extracts and Its Antioxidant and Antimicrobial Properties." *Particulate Science and Technology* 38 (8): 1019–26. https://doi.org/10.1080/02726351.2019.1658664.

He, Xiaojia, Hua Deng, and Huey min Hwang. 2019. "The Current Application of Nanotechnology in Food and Agriculture." *Journal of Food and Drug Analysis* 27 (1): 1–21. https://doi.org/10.1016/j.jfda.2018.12.002.

Holzinger, Michael, Alan Le Goff, and Serge Cosnier. 2014. "Nanomaterials for Biosensing Applications: A Review." *Frontiers in Chemistry* 2 (AUG): 1–10. https://doi.org/10.3389/fchem.2014.00063.

Huang, Yukun, Lei Mei, Xianggui Chen, and Qin Wang. 2018. "Recent Developments in Food Packaging Based on Nanomaterials." *Nanomaterials* 8 (10): 1–29. https://doi.org/10.3390/nano8100830.

Khan, Arif Ullah, Qipeng Yuan, Yun Wei, Shahab Ullah Khan, Kamran Tahir, Zia Ul Haq Khan, Aftab Ahmad, Farman Ali, Shafqat Ali, and Sadia Nazir. 2016. "Longan Fruit Juice Mediated Synthesis of Uniformly Dispersed Spherical AuNPs: Cytotoxicity against Human Breast Cancer Cell Line MCF-7, Antioxidant and Fluorescent Properties." *RSC Advances* 6 (28): 23775–82. https://doi.org/10.1039/c5ra27100b.

Kumar, Santosh, Jyotish Chandra Boro, Dharitri Ray, Avik Mukherjee, and Joydeep Dutta. 2019. "Bionanocomposite Films of Agar Incorporated with ZnO Nanoparticles as an Active Packaging Material for Shelf Life Extension of Green Grape." *Heliyon* 5 (6): e01867. https://doi.org/10.1016/j.heliyon.2019.e01867.

Kumar, Vinay, and Kavita Arora. 2020. "Trends in Nano-Inspired Biosensors for Plants." *Materials Science for Energy Technologies* 3: 255–73. https://doi.org/10.1016/j.mset.2019.10.004.

Mageswari, Anbazhagan, Ramachandran Srinivasan, Parthiban Subramanian, Nachimuthu Ramesh, and Kodiveri Muthukaliannan Gothandam. 2016. "Nanomaterials: Classification, Biological Synthesis and Characterization." In *Nanoscience in Food and Agriculture*, edited by Shivendu Ranjan, Nandita Dasgupta, and Eric Lichtfouse, 23:31–71. Sustainable Agriculture Reviews. Cham: Springer International Publishing. https://doi.org/10.1007/978-3-319-48009-1_2.

Mittal, Amit Kumar, Yusuf Chisti, and Uttam Chand Banerjee. 2013. "Synthesis of Metallic Nanoparticles Using Plant Extracts." *Biotechnology Advances* 31 (2): 346–56. https://doi.org/10.1016/j.biotechadv.2013.01.003.

Mohaghegh, Seraj, Karim Osouli-Bostanabad, Hossein Nazemiyeh, Yousef Javadzadeh, Alireza Parvizpur, Mohammad Barzegar-Jalali, and Khosro Adibkia. 2020. "A Comparative Study of Eco-Friendly Silver Nanoparticles Synthesis Using Prunus Domestica Plum Extract and Sodium Citrate as Reducing Agents." *Advanced Powder Technology* 31 (3): 1169–80. https://doi.org/10.1016/j.apt.2019.12.039.

Naseem, Taiba, and Tayyiba Durrani. 2021. "The Role of Some Important Metal Oxide Nanoparticles for Wastewater and Antibacterial Applications: A Review." *Environmental Chemistry and Ecotoxicology* 3: 59–75.

https://doi.org/10.1016/j.enceco.2020.12.001.

Oves, Mohammad, Mohammad Zain Khan, and Iqbal M.I. Ismail. 2017. *Modern Age Environmental Problems and Their Remediation*. Edited by Mohammad Oves, Mohammad Zain Khan, and Iqbal M.I. Ismail. *Modern Age Environmental Problems and Their Remediation*. Cham: Springer International Publishing. https://doi.org/10.1007/978-3-319-64501-8.

Patra, Jayanta Kumar, Gitishree Das, Leonardo Fernandes Fraceto, Estefania Vangelie Ramos Campos, Maria Del Pilar Rodriguez-Torres, Laura Susana Acosta-Torres, Luis Armando Diaz-Torres, et al. 2018. "Nano Based Drug Delivery Systems: Recent Developments and Future Prospects." *Journal of Nanobiotechnology* 16 (1): 71. https://doi.org/10.1186/s12951-018-0392-8.

Pozdnyakov, Alexander S., Artem I. Emel'yanov, Nadezhda P. Kuznetsova, Tamara G. Ermakova, Tat'yana V. Fadeeva, Larisa M. Sosedova, and Galina F. Prozorova. 2016. "Nontoxic Hydrophilic Polymeric Nanocomposites Containing Silver Nanoparticles with Strong Antimicrobial Activity." *International Journal of Nanomedicine* 11: 1295–1304. https://doi.org/10.2147/IJN.S98995.

R.D. Singh, Rajni Garg. 2011. *Inorganic Chemistry - 1st Edition. McGraw Hill Publications*. https://www.expresslibrary.mheducation.com/product/inorganic-chemistry50059658?fbclid=IwAR2qcTkJx5DV85LViODqWmjJ8UpLxnXsZoTBd590o rV2NDUFTVI_ZNtmRhs.

Rai, Vineeta, Sefali Acharya, and Nrisingha Dey. 2012. "Implications of Nanobiosensors in Agriculture." *Journal of Biomaterials and Nanobiotechnology* 03 (02): 315–24. https://doi.org/10.4236/jbnb.2012.322039.

Rajni Garg. 2011. "Effect of Addition of Crown Ether on the Micellar Behavior of

Dodecyltrimethylammonium Chloride in Aqueous Media." *Grin Verlag.*

Rajni Garg. 2012. *Supramolecular Chemistry of Host-Guest Complexes. Munich, GRIN Verlag.* https://www.grin.com/document/187894.

Sharma, Rajat, Rajni Garg, and Avnesh Kumari. 2020. "A Review on Biogenic Synthesis, Applications and Toxicity Aspects of Zinc Oxide Nanoparticles." *EXCLI Journal.* https://doi.org/10.17179/excli2020-2842.

Shukla, Abhay, Rakesh Kumar, Javed Mazher, and Adrian Balan. 2009. "Graphene Made Easy: High Quality, Large-Area Samples." *Solid State Communications* 149 (17–18): 718–21. https://doi.org/10.1016/j.ssc.2009.02.007.

Sozer, Nesli, and Jozef L. Kokini. 2009. "Nanotechnology and Its Applications in the Food Sector." *Trends in Biotechnology* 27 (2): 82–89. https://doi.org/10.1016/j.tibtech.2008.10.010.

Vazquez-Munoz, Roberto, and Jose L. Lopez-Ribot. 2020. "Nanotechnology as an Alternative to Reduce the Spread of COVID-19." *Challenges* 11 (2): 15. https://doi.org/10.3390/challe11020015.

Vipan, Bal, Chander Mahajan, Ritu Tandon, Swati Kapoor, and Mohinder Kaur Sidhu. 2018. "Natural Coatings for Shelf-Life Enhancement and Quality Maintenance of Fresh Fruits and Vegetables-A Review." *Journal of Postharvest Technology* 06 (1): 12–26. http://www.jpht.info.

Zhou, Yazhou, Jipei Huang, Weidong Shi, Yi Li, Yunyan Wu, Qinqin Liu, Jia Zhu, et al. 2018. "Ecofriendly and Environment-Friendly Synthesis of Size-Controlled Silver Nanoparticles/Graphene Composites for Antimicrobial and SERS Actions." *Applied Surface Science* 457: 1000–1008. https://doi.org/10.1016/j.apsusc.2018.07.040.